SCIENCE AROUND US

Worms

By Peter Murray

THE CHILD'S WORLD®
CHANHASSEN, MINNESOTA

Published in the United States of America by The Child's World®
PO Box 326, Chanhassen, MN 55317-0326
800-599-READ
www.childsworld.com

Content Advisers:
Jim Rising, PhD,
Professor of Zoology,
University of Toronto,
Department of Zoology,
Toronto, Ontario,
Canada, and Trudy
Rising, Educational
Consultant, Toronto,
Ontario, Canada

Photo Credits:
Cover/frontispiece: Getty Images/Photodisc/Ryan McVay; cover corner: Corbis.

Interior: Animals Animals/Earth Scenes: 4 (Wild & Natural), 5 (Marian Bacon), 11 (James Robinson), 25 (John Anderson); Dr. Jeremy Burgess/Photo Researchers: 18; Corbis: 6 (Jeffrey L. Rotman), 8 (Michael & Patricia Fogden), 12 (Lester V. Bergman), 15 (Douglas Peebles), 17 (AFP), 19 (Michael Boys), 20 (Clouds Hill Imaging Ltd.), 24 (Lawson Wood), 29 (Ralph White); E. R. Degginger/Dembinsky Photo Associates: 21, 27; David M. Dennis/Animals Animals/Earth Scenes: 10, 23; OSF/Animals Animals/Earth Scenes: 14 (Kathy Atkinson), 16 (Alistair MacEwen).

The Child's World®: Mary Berendes, Publishing Director

Editorial Directions, Inc.: E. Russell Primm, Editorial Director; Pam Rosenberg, Line Editor; Katie Marsico, Assistant Editor; Matt Messbarger, Editorial Assistant; Susan Hindman, Copy Editor; Susan Ashley, Proofreader; Peter Garnham, Terry Johnson, Olivia Nellums, Katherine Trickle, and Stephen Carl Wender, Fact Checkers; Tim Griffin/IndexServ, Indexer; Cian Loughlin O'Day, Photo Researcher; Linda S. Koutris, Photo Selector

The Design Lab: Kathleen Petelinsek, Design and Page Production

Copyright © 2005 by The Child's World®
All rights reserved. No part of this book may be reproduced or utilized in any form or by any means without written permission from the publisher.

Library of Congress Cataloging-in-Publication Data
Murray, Peter.
   Worms / by Peter Murray.
      p. cm. — (Science around us)
   Includes bibliographical references and index.
   ISBN 1-59296-276-9 (lib. bd. : alk. paper) 1. Worms—Juvenile literature. [1. Worms.]
I. Title. II. Science around us (Child's World (Firm))
   QL386.6.M87 2004
   592'.3—dc22   2003027221

Table of Contents

CHAPTER ONE
4 A Simple Tube

CHAPTER TWO
6 Flatworms

CHAPTER THREE
13 Ribbon Worms and Rotifers

CHAPTER FOUR
17 Roundworms

CHAPTER FIVE
21 Segmented Worms

28 Glossary

29 Did You Know?

30 The Animal Kingdom

31 How to Learn More about Worms

32 Index

CHAPTER ONE

A Simple Tube

What do you have in common with an earthworm? More than you might think. You have a heart, blood, and skin. So do earthworms. You get hungry, and you eat. So do worms.

There are differences, of course. Worms have no arms or legs. They have no bones. They can't talk or ride a bike or throw a ball.

You are not an earthworm, but you are built around the same

An earthworm is shaped like a tube and has no arms or legs.

simple plan: Your body is a tube. Food goes in one end, and waste comes out the other.

There are hundreds of thousands of different **species** of worms living on every part of our planet, from the deepest oceans to the highest mountains. A single handful of dirt might contain more than 100,000 microscopic worms. Some worms live inside the bodies of other animals, including human beings.

Worms can be divided into three basic groups: flatworms, roundworms, and segmented worms.

A flatworm swims in the Red Sea. Worms live on every part of our planet, including the oceans.

CHAPTER TWO

Flatworms

As their name suggests, flatworms have flattened bodies. These primitive worms were the first animals to have bilateral symmetry. This means that instead of being shapeless blobs or being

A colorful flatworm grazes on a reef in Australia. A flatworm has one opening in the center of its body that it uses for taking in food and letting out waste.

round like a wheel, the worms have left and right sides that are mirror images of each other. They have heads and tails, and the ability to move. Flatworms were among the first animals to develop muscles. While other primitive creatures were simply tossed about by the ocean currents, worms could swim or creep along the bottom of the ocean.

Flatworms do not have lungs, hearts, or blood. They absorb oxygen through their thin skin. Unlike the more advanced worms, flatworms are not tube-shaped—single openings in the center of their bodies are used for both eating and waste disposal.

Many flatworms are **parasitic.** These parasites live inside the bodies of other animals.

Scientists estimate that there are more than 20,000 species of flatworms. They are divided into three classes: turbellarians, flukes, and tapeworms.

TURBELLARIANS

These flatworms live in the oceans, in freshwater, and in moist soil. They are able to move through the water and over surfaces by wriggling their bodies. Turbellarians have simple brains and tiny eyespots that detect light and movement.

> Planarians are flatworms that are often used in school science experiments. If you cut the head of a planarian in half lengthwise, the worm will soon grow two complete heads!

Turbellarians that live on land need moist soil to survive. This striped turbellarian was found in a rain forest in Costa Rica.

Most turbellarians are small and drab, but some ocean species, known as polyclads, can be brightly colored and up to 50 centimeters (20 inches) long. Their bright colors mimic those of the poisonous sea slug and are a warning to **predators:** Don't eat me. I might not taste good and might even make you sick!

FLUKES

Flukes are small parasitic flatworms that live in the bodies of more advanced animals. Most flukes are quite small, but when they invade a **host** body, they can make big problems.

There are an estimated 12,000 species of flukes. Every species infects different animals in different ways. Blood flukes live in the bloodstream. Liver flukes make their way to the liver and **bile ducts** of sheep, humans, and other mammals, where they can cause serious illness and even death.

The human liver fluke has a **complex** life cycle. Its eggs pass out of its human host through the digestive tract. If the eggs reach a lake or stream, they are eaten by freshwater snails. The eggs hatch inside the snail, feed on the snail's tissues, and eventually emerge as free-swimming **larvae.** The larvae then invade the body of a passing fish. If the uncooked fish is eaten by humans, the young flukes crawl from the intestines into the liver, attach their **suckers,** and begin to lay eggs. Hundreds of millions of people worldwide suffer from the painful symptoms caused by flukes living inside them.

The Chinese liver fluke causes sickness in humans. People become infected by eating fish that is raw or undercooked.

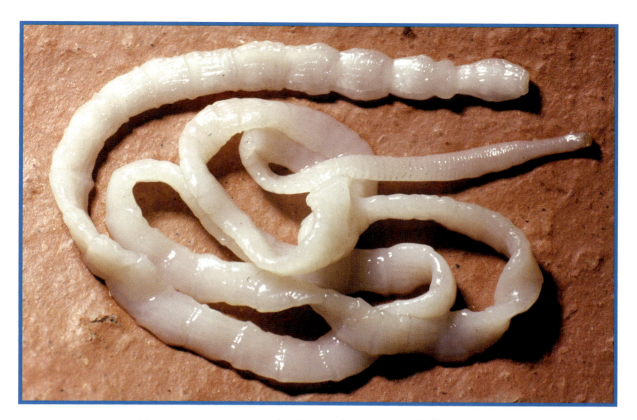

Tapeworms are parasites that live in the intestines of other animals, including humans. This tapeworm was removed from a cat.

TAPEWORMS

The tapeworm lives in the intestines of other animals. It lets its host do all the eating and digesting—the worm simply absorbs nutrients through its body walls.

A tapeworm has a specialized head called a scolex. As the tapeworm feeds, flat segments grow out from the scolex, like inches on a tape measure. Each small segment contains thousands of eggs. The worm keeps adding segments, growing longer and longer.

In time, the segments break away. They are carried out of the host's body in the waste. Eventually, some of those eggs find their way into other animals, and the cycle repeats itself.

> **The tapeworm's scolex has tiny suckers and hooks that attach firmly to the inside wall of the host animal's intestine.**

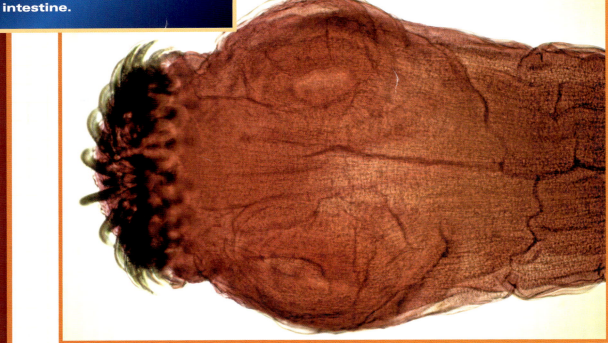

The scolex of the beef tapeworm (above) attaches itself to the wall of a human's intestines. People can become infected with the tapeworm by eating undercooked beef.

CHAPTER THREE

Ribbon Worms and Rotifers

RIBBON WORMS

If you live near the ocean, you may have seen brightly colored ribbon worms hiding under rocks and among strands of seaweed. There are about 900 species of ribbon worms. They can be pink, white, purple, green, orange, or even striped.

Ribbon worms were probably the first animals to **evolve** one-way digestive tracts. Food goes in one end and comes out the other. This means that a ribbon worm never has to stop eating. It can keep eating and growing and growing and eating. Ribbon worms also have simple circulatory systems—red blood cells carry oxygen throughout their bodies.

> Most ribbon worms live in salt water.

This ribbon worm from Australia is just one of nearly 1,000 kinds of ribbon worms that have been identified.

The longest creature on earth is a type of ribbon worm that grows up to 55 meters (180 feet) long.

Many ribbon worm species have sharp tubes in their mouths that can shoot out to spear tiny shrimp, smaller worms, or other prey. The tubes contain a poison strong enough to paralyze their victims.

ROTIFERS

Rotifers live just about everywhere there is moisture. They are found frozen in the Arctic ice. They live in moist dirt and on the trunks of trees. Even a puddle left after a rain is full of rotifers. They are one of our most common worms—but you won't see one without a microscope. Most rotifers are too tiny to be seen with the naked eye.

There may be thousands of rotifers living on this tree in the Hoh Rain Forest in Olympic National Park in the state of Washington. But don't try to find them— you can't see them without a microscope!

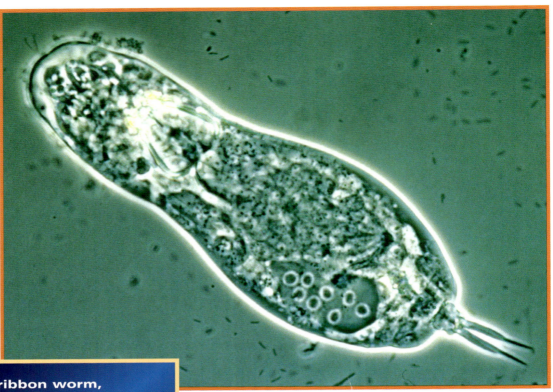

Most rotifers are transparent—you can see their insides right through their skin!

Like a ribbon worm, a rotifer has a one-way digestive system—it takes in food through a mouth at the large end and expels waste through an opening—called an anus—at its pointed end.

Rotifers are named for two rotating "wheels" of tiny hairs, or cilia, that grow at the rotifer's mouth end. By waving its cilia in a circular pattern, the rotifer swirls the water and draws food into its mouth. It can also use its cilia like a propeller to move from one place to another.

CHAPTER FOUR

Roundworms

No one knows how many species of roundworms there are on earth. Twelve thousand have been named, but there are probably hundreds of thousands that remain to be discovered.

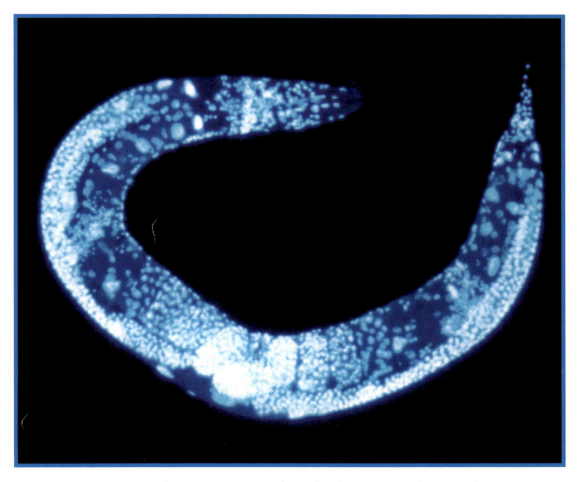

Many roundworms are parasites that infect dogs, cats, or other animals.

As you can see from this roundworm wrapped around a blade of grass, not all the members of this worm group are small.

Roundworms are smooth-skinned, round, and usually pointed at both ends. Like ribbon worms, they have one-way digestive systems and blood that moves oxygen to all their body parts. Roundworms can be up to 9 meters (30 feet) long, but most roundworms are very small. One acre of garden soil might contain 5 billion tiny roundworms!

The soil in this garden probably contains millions of roundworms.

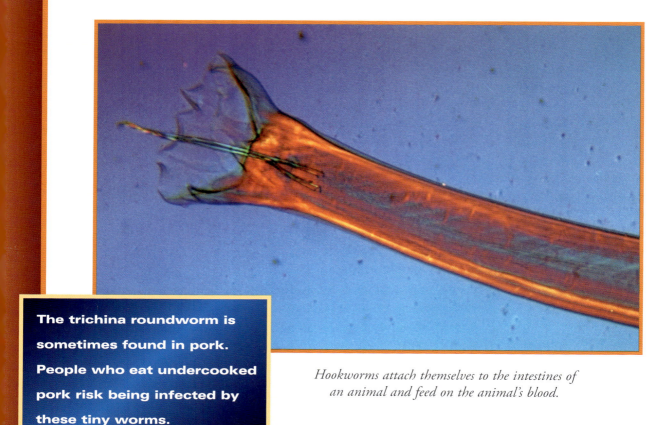

Hookworms attach themselves to the intestines of an animal and feed on the animal's blood.

The trichina roundworm is sometimes found in pork. People who eat undercooked pork risk being infected by these tiny worms.

Most roundworms are free-living—they eat decaying vegetable matter. Some roundworms are parasites. Heartworms are a type of roundworm that can cause disease in dogs. Several types of roundworms, including hookworms and pinworms, are human parasites. The human intestinal roundworm, which can grow up to 40 centimeters (16 inches) long, infects about 800 million people worldwide.

CHAPTER FIVE

SEGMENTED WORMS

Segmented worms, known as annelids, are complex animals. Annelids are more closely related to insects than they are to the other worms.

An annelid's body is made up of a series of rings, called segments. This segmented body allows the worm to twist and turn and stretch

Look closely and you will see the rings or segments that make up this earthworm's body.

itself in any direction. Each segment has several tiny **bristles** that work like feet. These bristles help the worm crawl or swim.

Inside the bodies of many annelids are several organs, including two stomachs, a gizzard for grinding food, a simple brain, and more than one heart. Most annelids live in soil or the sediment at the bottom of the ocean.

EARTHWORMS

Earthworms are among our most valuable worms. They are found in just about every type of soil, where they eat leaves and other decaying vegetable matter. Once the leaves are digested, the earthworm deposits its waste on the surface. These tiny waste piles are called worm castings. Castings help enrich the soil for new plant life.

There are more than 2,000 species of earthworms. They are a favorite food of many birds and other animals, and they need to stay moist. Because of this, earthworms spend their days underground. At night, they come to the surface to eat leaves or to **mate.**

> One acre (.405 hectares) of healthy garden soil might contain 900 kilograms (1,984 pounds) of earthworms—enough to fill the back of a pickup truck!

Nightcrawlers and other kinds of earthworms help keep the soil healthy so that plants can grow.

BRISTLEWORMS

True to their name, bristleworms have a lot of bristles. Some look like centipedes or fuzzy caterpillars.

Most of the 8,000 species of bristleworms live in the oceans. Ragworms live in **burrows** in the seabed. They are good swimmers with eyes and strong jaws. Ragworms will chase down smaller animals and eat them. Fireworms live in coral reefs near the sponges and coral that they feed upon. They are called fireworms because they can make their bodies glow in the dark, like fireflies. Instead, they usually stay in one place. Parchment worms spend most of their lives buried in

If you see a fireworm, don't touch it. If you do, the poison on its bristles will make your skin feel like it is burning.

The tentacles of a group of feather-duster worms stick out from tubes made from mucus and sand.

The bristles of some brightly colored fireworms are poisonous.

U-shaped burrows in the seabed. They stir the water at the burrow entrance to draw in food particles.

Feather-duster worms and fanworms live inside hard, protective tubes they build from mucus and sand. They hide inside these tubes with their feeding **tentacles** exposed like the feathers of a dust mop. When danger threatens, the worms quickly draw back into their tubes.

LEECHES

> **Some aquatic leeches eat snails, larvae, and worms.**

Leeches are parasitic annelids. They have no bristles, and they swim or crawl by undulating their bodies. Leeches live by sucking blood from fish and other animals. Once a leech attaches itself to a host with its sucker, it pierces the skin and slowly draws blood into its gut. One feeding is enough for a leech to live on for several months.

Leeches were once commonly used by doctors. It was believed that leeches could draw the "bad blood" from someone who was very sick. A doctor might attach several leeches at a time to a patient and let them drink their fill.

◆ ◆ ◆

For half a billion years, worms have been crawling in and around our planet. Most worms are invisible and harmless to humans. They are hiding underground or deep in the ocean, going about their wormy business. Although some worms can cause illness in humans, others enrich our soil and provide food for other living things. Like all creatures, worms play an important role in the balance of nature.

When a leech bites a human, its saliva keeps the blood from clotting so that it will flow more quickly. Today, some doctors use leeches to help keep blood flowing after delicate operations to reattach fingers, toes, and other body parts.

GLOSSARY

aquatic (uh-KWOT-ik) If an animal is aquatic, it lives in water.

bile ducts (BILE DUCKTS) Bile ducts are tubes that carry bile, a green liquid made by the liver, to the intestines, where it helps aid in the process of digestion.

bristles (BRISS-uhlz) Bristles are short, stiff hairs.

burrows (BUR-ohs) Burrows are tunnels or holes in the ground made by animals that use them for shelter.

complex (kuhm-PLEKS) If something is complex, it is complicated or has many different parts or steps in a process.

evolve (ih-VOLV) To evolve means to change slowly over time.

host (HOHST) A host is a plant or animal that is used by a parasite for nutrition.

larvae (LAR-vee) Larvae are insects at the stage of development between egg and pupa when they look like worms.

mate (MAYT) If two living things mate, they join together to produce young.

parasitic (pa-ruh-SIT-ik) An animal or plant that is parasitic gets its food by living on or inside another living thing.

predators (PRED-uh-turz) Predators are animals that hunt other animals for food.

species (SPEE-sheez) A species is a certain type of living thing. Worms of the same species can mate and produce young. Worms of different species cannot produce young together.

suckers (SUHK-urz) Suckers are body parts of certain animals that allow the animals to attach themselves to surfaces.

tentacles (TEN-tuh-kuhls) Tentacles are long, flexible limbs that can be used for grabbing and moving.

DID YOU KNOW?

- Tapeworms living in whale intestines can grow to more than 30 meters (100 feet) long.

- Giant Australian earthworms can grow to more than 3.5 meters (12 feet) long.

- The bristleworm known as the sea mouse can grow up to 20 centimeters (8 inches) long.

- Giant tube worms more than 1 meter (3 feet) long have been found growing near vents in the ocean floor, about 2.5 kilometers (1.5 miles) beneath the surface. The vents supply heat and gases that the worms need to live and grow.

- Usually the smaller the worm, the shorter its life span. Some very small worms may live only 45 to 60 days, while some of the largest worms can live to be more than 50 years old.

- Earthworms will drown if they remain underground during a heavy rainfall. That is why you see so many earthworms on the surface after a rainstorm.

- The scientific name of the longest worm known is *Lineus longissimus*. These worms are commonly more than 4.5 meters (15 feet) long, and some of them have been reported to be more than 45 meters (150 feet) in length!

Giant tube worms grow very quickly—scientists have observed some growing in length at a rate of 84 centimeters (33 inches) per year.

THE ANIMAL KINGDOM

VERTEBRATES

fish

amphibians

reptiles

birds

mammals

INVERTEBRATES

sponges

worms

insects

spiders & scorpions

mollusks & crustaceans

sea stars

sea jellies

HOW TO LEARN MORE ABOUT WORMS

At the Library

Cronin, Doreen, and Harry Bliss (illustrator). *Diary of an Earthworm.*
New York: Joanna Cotler Books, 2003.

Dell'Oro, Susanne Paul. *Tunneling Earthworms.*
Minneapolis: Lerner Publications, 2001.

Himmelman, John. *An Earthworm's Life.*
Danbury, Conn.: Children's Press, 2000.

On the Web

VISIT OUR HOME PAGE FOR LOTS OF LINKS ABOUT WORMS:
http://www.childsworld.com/links.html
Note to Parents, Teachers, and Librarians: We routinely check our Web links to make sure they're safe, active sites—so encourage your readers to check them out!

Places to Visit or Contact

BROOKLYN BOTANIC GARDEN
To learn about how worms can be used in composting
1000 Washington Avenue
Brooklyn, NY 11225
718/623-7200

WORM DIGEST
*To write for more information about worms
and how they help our environment*
PO Box 544
Eugene, OR 97440-0544
541/485-0456

INDEX

annelids, 21–26
anus, 16

bilateral symmetry, 6–7
blood flukes, 9
bristles, 22
bristleworms, 24–26

castings, 22
cilia, 16
circulatory systems, 13, 22

digestive systems, 13, 16, 19, 22

earthworms, 4, *4*, *21*, 22–23, *23*
eggs, 10, 12

feather-duster worms, 25–26, *25*
fireworms, 24, *24*, 25
flatworms, 5, 6–12, *6*
flukes, 7, 9–10, *10*

groups, 5

heartworms, 20
Hoh Rain Forest, *15*
hookworms, 20, *20*

leeches, 26, *27*
liver flukes, 9, 10, *10*

muscles, 7

nightcrawlers, *23*

Olympic National Park, *15*
oxygen, 7, 19

parasitic worms, 7, 20, 26
parchment worms, 24–25
pinworms, 20
planarians, 8
poison, 14, *24*, 25
polyclads, 9

ragworms, 24
ribbon worms, 13–14, *14*
rotifers, 15–16, *15*, *16*
roundworms, 5, 17, *17*, *18*, 19–20, *19*

scolex, 11–12, *12*
segmented worms, 5, 21–26
segments, 21–22
skin, 7, 19
species, 5, 7, 9, 13, 17, 22, 24

tapeworms, 7, 11–12, *11*, *12*
tentacles, 25
trichina roundworms, 20
turbellarians, 7, 8–9, *8*

worm castings, 22

About the Author

Peter Murray has written more than 80 children's books on science, nature, history, and other topics. An animal lover, Pete lives in Golden Valley, Minnesota, in a house with one woman, two poodles, several dozen spiders, thousands of microscopic dust mites, and an occasional mouse.